Foreword

The modern laboratory is a hub of innovation and discovery, a place where groundbreaking research and technological advancements are born. However, alongside the pursuit of knowledge comes the paramount responsibility of ensuring the safety and well-being of those who work within these environments. "Laboratory Health & Safety: A Comprehensive Guide" by David Ojowu provides an invaluable resource for achieving this balance.

In my many years of working in the fields of biochemistry and biotechnology, I have witnessed firsthand the critical importance of adhering to stringent health and safety protocols. Laboratories are complex ecosystems, often housing hazardous materials and sophisticated equipment that, if mishandled, can pose significant risks. This comprehensive guide not only underscores the necessity of safety but also offers practical, actionable strategies to mitigate potential hazards.

David Ojowu, with their extensive academic background in biochemistry (BSc) and biotechnology (MSc), brings a wealth of knowledge and experience to this subject. Their meticulous approach to detail and commitment to fostering a safe laboratory environment are evident throughout the pages of this book. From fundamental principles to advanced safety techniques, every chapter is designed to equip laboratory personnel with the tools they need to work safely and efficiently.

This guide is particularly timely, given the increasing complexity of contemporary research and the evolving regulatory landscape. As we push the boundaries of scientific inquiry, the safety of our researchers and staff must remain a top priority. "Laboratory Health & Safety: A Comprehensive Guide" serves as an essential manual for educators, students, researchers, and laboratory managers alike.

I am confident that this book will not only enhance the safety practices within your laboratory but also inspire a culture of vigilance and responsibility. It is a must-read for anyone committed to the pursuit of scientific excellence within a safe and secure environment.

Dr Dwight Hardy
Senior Scientist

Preface

The journey of scientific discovery is both exhilarating and challenging, marked by the balance between curiosity and caution. As a student, I found myself navigating the complexities of laboratory work with a sense of wonder but also with a lack of comprehensive guidance. My early experiences in biochemistry and biotechnology were characterised by a trial-and-error approach, where access to detailed health and safety protocols was limited. This gap left me feeling unprepared and, at times, vulnerable.

Years later, as both a supply teacher and a professional in the biotechnology industry, I witnessed similar struggles among my students and colleagues. The educational system, while rich in theoretical knowledge, often fell short in providing practical, hands-on training that reflects the realities of modern laboratory environments. It became evident that there was a disconnect between what was being taught and the actual demands of the field. The lack of emphasis on up-to-date safety practices was a glaring omission that needed to be addressed.

In the industry, I observed that lapses in safety protocols were not confined to the classroom. Many workplaces operated without the benefit of comprehensive safety training, leading to avoidable accidents and inefficiencies. The fast-paced nature of industrial research and production often overshadowed the critical importance of maintaining rigorous safety standards.

Driven by these experiences, I embarked on a mission to bridge this gap. "Laboratory Health & Safety: A Comprehensive Guide" is the culmination of my efforts to create a resource that I wish I had access to earlier. This book is designed to provide a thorough, practical understanding of laboratory safety, equipping both students and professionals with the knowledge they need to work safely and effectively in academic and industrial settings.

My motivation is deeply rooted in the desire to have a personal impact on this field. I believe that by enhancing safety standards and education, we can foster a culture of responsibility and excellence in our laboratories. This book is not just a guide; it is a reflection of my commitment to improving the educational system and industry practices, adapting them to meet the demands of our current day and reality.

I hope that this comprehensive guide will serve as a valuable tool for students, educators, researchers, and industry professionals alike. By integrating up-to-date safety practices into everyday laboratory work, we can create environments where scientific innovation can thrive without compromising on safety. It is my sincere hope that this book will inspire and empower a new generation of scientists and professionals to pursue their work with confidence and caution.

Abstract

In everything we do, prioritising health and safety is paramount. Whether in everyday tasks or complex scientific work, ensuring a safe environment protects everyone involved and prevents accidents and injuries. This book, "Laboratory Health & Safety: A Comprehensive Guide," focuses on laboratory health and safety, providing essential guidelines and best practices to keep our laboratories safe and efficient.

From basic safety protocols to advanced hazard management techniques, this comprehensive guide covers a wide array of topics crucial for maintaining a safe laboratory environment. The book is designed for a diverse audience, including seasoned scientists, educators, industry professionals, and curious newcomers. Each chapter delves into specific aspects of laboratory safety, offering practical tips and actionable advice to prevent accidents and minimise health risks.

The chapters include detailed discussions on proper handling and storage of hazardous materials, emergency response procedures, personal protective equipment (PPE) usage, and the importance of regular safety audits. Real-world case studies and examples illustrate the potential consequences of neglecting safety protocols, emphasising the critical need for vigilance and adherence to best practices.

The motivation for this book stems from the author's own experiences as a student and teacher, where the lack of comprehensive safety information often posed challenges. By addressing these gaps, the book aims to bridge the disconnect between theoretical knowledge and practical application, both in academia and industry.

Ultimately, this guide serves as a vital resource for anyone working in a laboratory setting, ensuring that safety is never compromised in the pursuit of scientific excellence. By following the guidelines and best practices outlined in this book, readers will be well-equipped to create and maintain a safe and productive laboratory environment.

Glossary

Acrylamide:
A chemical used in gel electrophoresis that is toxic and a potential carcinogen. Proper handling and disposal are essential to avoid health risks.

Autoclave:
A device used to sterilise equipment and materials by applying high-pressure saturated steam. Essential for preventing contamination in laboratory work.

Biosafety Cabinet (BSC):
An enclosed, ventilated laboratory workspace for safely working with materials contaminated with pathogens. Provides a sterile environment for handling hazardous substances.

Chemical Hygiene Plan (CHP):
A written program developed and implemented by an employer that sets forth procedures, equipment, personal protective equipment, and work practices to protect employees from the health hazards presented by hazardous chemicals used in that particular workplace.

Ergonomics:
The study of people's efficiency in their working environment, often applied to ensure that laboratory workstations are designed to minimise discomfort and risk of injury.

Fume Hood:
A ventilated enclosure in a chemistry laboratory, in which harmful volatile chemicals can be used or kept. Protects users from inhaling toxic gases.

Hazardous Waste:
Waste materials that are dangerous or potentially harmful to human health or the environment. Requires special handling, storage, and disposal procedures.

Material Safety Data Sheet (MSDS):
A document that provides detailed information on the properties of a chemical, including its hazards, safe handling practices, and emergency control measures.

Personal Protective Equipment (PPE):
Equipment worn to minimise exposure to hazards that cause serious workplace injuries and illnesses. Includes items such as gloves, safety glasses, and lab coats.

Polymerase Chain Reaction (PCR):
A technique used in molecular biology to amplify a single copy or a few copies of a segment of DNA across several orders of magnitude, generating thousands to millions of copies of a particular DNA sequence.

Standard Operating Procedure (SOP):
A set of step-by-step instructions compiled by an organisation to help workers carry out complex routine operations. Ensures consistency and compliance with safety standards.

Sterilisation:
The process of making something free from bacteria or other living microorganisms. Essential in preventing contamination in laboratory experiments.

Titration:
A technique where a solution of known concentration is used to determine the concentration of an unknown solution. Commonly used in chemical analysis.

Ultraviolet (UV) Light:
A type of electromagnetic radiation used in laboratories for various purposes, including sterilisation and detecting certain substances. Prolonged exposure can be harmful.

Ventilation:
The process of supplying fresh air and removing contaminated air from a space. Critical in laboratories to ensure a safe working environment.

Zoonosis:
A disease that can be transmitted from animals to humans. Laboratory safety protocols often include measures to prevent zoonotic infections.

CHAPTER 1

Introduction

In everything we do, prioritising health and safety is paramount. Whether in everyday tasks or complex scientific work, ensuring a safe environment protects everyone involved and prevents accidents and injuries. This book focuses on laboratory health and safety, providing essential guidelines and best practices to keep our laboratories safe and efficient. Whether you're a seasoned scientist or a curious newcomer, these tips will help keep you safe and sound in the laboratory. Following the wrong practices can lead to accidents, injuries, or even serious health risks, so let's make sure we do it right!

CHAPTER 2

ISO: International Organisation for Standardisation

Introduction to ISO Standards

ISO standards relevant to laboratory settings are crucial for maintaining consistency, safety, and quality across laboratories worldwide. These standards create a global benchmark for excellence. Image: Animated ISO logo, globe showing interconnected laboratories.

Key ISO Standards: Definition & Explanation

Key standards include ISO 15189 for medical laboratories, ISO 17025 for testing and calibration laboratories, and ISO 45001 for occupational health and safety.

ISO 15189: Focuses on quality management, ensuring that laboratory personnel are competent and equipment is properly calibrated.

ISO 17025: Specifies requirements for testing and calibration, ensuring accurate results and reliable reporting.

ISO 45001: Emphasises hazard identification, risk assessment, and emergency preparedness.

Image: Animated checklist of ISO standards with laboratories meeting each requirement.

Importance of ISO Accreditation

Every laboratory should strive for ISO accreditation. Accreditation verifies that your laboratory meets international safety and quality standards, ensuring reliable results and safe practices. It also enhances credibility, improves efficiency, and helps in compliance with legal and regulatory requirements (International Organisation for Standardisation, 2021).

Image: Laboratories applying for accreditation, officials reviewing laboratory processes, certificate being awarded.

CHAPTER 3

Biosafety Levels

Importance of Biosafety Levels

Biosafety levels are crucial for handling biological agents safely. Each level has specific safety measures to protect laboratory personnel and the environment. Image: Various laboratory settings with different safety measures.

BSL-1: Definition & Explanation

BSL-1 involves handling agents that are not known to cause disease in healthy humans. Basic safety measures are required

Do :
- Wear gloves, a lab coat, and safety glasses.
- Wash hands after handling materials.

Don't :
- Handle materials carelessly or without proper attire.
- Ignore hand hygiene.

Image: Split screen with correct practice on one side and incorrect on the other.

BSL-2: Definition & Explanation

BSL-2 involves moderate-risk agents that pose a danger through accidental ingestion or contact with mucous membranes, such as Salmonella, Hepatitis B virus, or human blood.

Do :
· Use personal protective equipment (PPE) including gloves, lab coat, and face shield.
· Follow protocols such as not touching the face, washing hands, and properly disposing of sharps in

Don't :
· Ignore PPE.
· Eat or drink in the laboratory.
· Improperly dispose of sharps

Image: Split screen with correct practice on one side and incorrect on the other.

BSL-3: Definition & Explanation

BSL-3 involves agents that can cause serious or potentially lethal diseases through inhalation, such as Mycobacterium tuberculosis, SARS-CoV-2, or West Nile virus.

Do : · Use biosafety cabinets and wear respiratory protection.
· Follow decontamination protocols for equipment and surfaces.

Don't : · Work without proper gear or respiratory protection.
· Neglect decontamination protocols.

Image: Proper containment practices

BSL-4: Definition & Explanation

BSL-4 is the highest level, dealing with dangerous and exotic agents that pose a high risk of life-threatening disease, such as Ebola, Marburg viruses, or Lassa fever.

Do : · Wear a full-body suit with positive pressure.
- Follow strict entry and exit procedures.

Don't : · Bypass security barriers.
- Use standard laboratory equipment for BSL-4 agents.

Image: Extreme containment measures vs. casual attire.

CHAPTER 4

General Safety Precautions

Dress code

Appropriate laboratory attire is essential to protect you from chemical spills, splashes, and other hazards.

Do : · Wear a lab coat, gloves, goggles, and closed-toe shoes.
· Tie back long hair and avoid loose clothing.

Don't : · Wear shorts or open-toed shoes.
· Have loose hair and clothing in the laboratory.

Image: Proper vs. improper attire.

Laboratory Hygiene

Maintaining laboratory hygiene is crucial for preventing contamination and ensuring a safe working environment.

Do :
- Disinfect work surfaces regularly.
- Properly dispose of waste.
- Wash hands after handling hazardous materials.

Don't :
- Leave spills uncleaned.
- Improperly dispose of waste.
- Neglect hand hygiene.

Image: Clean vs. dirty laboratory workstations

Risk Assessment

Risk assessments help identify potential hazards and implement control measures to prevent accidents.

Do : · Review a checklist and identify hazards such as chemical spills, biological agents, and fire risks.
· Implement controls like fume hoods for volatile substances, biohazard cabinets for pathogens, and fire extinguishers.

Don't : · Start experiments without any checks.
· Use chemicals near open flames.
· Handle pathogens without proper containment.

Image: Risk assessment forms, hazardous materials, fume hoods, biohazard cabinets, fire extinguishers.

Laboratory Organisation

Organising the laboratory efficiently ensures a smooth workflow and minimises the risk of accidents.

Do : · Label chemicals.
  · Store equipment in designated areas.
  · Maintain clear workspaces.

Don't : · Leave chemicals unlabeled.
  · Place equipment haphazardly.
  · Clutter the workspace.

Image: of an organised laboratory

Workflow Management

Efficient workflow management in the laboratory increases productivity and safety.

Do : · Plan experiments.
    · Follow standard operating procedures (SOPs).
    · Document results meticulously.

Don't : · Improvise experiments.
    · Ignore SOPs.
    · Fail to record data.

Image: of an organised laboratory

Emergency Procedures

Knowing emergency procedures is vital for responding quickly and effectively to accidents.

Do : · Locate and understand the use of safety showers, eyewash stations, first aid kits, and emergency exits.
    · Practise emergency drills.

Don't : · Be unaware of emergency equipment locations.
    · Ignore safety drills.
    · Improperly handle emergency situations.

CHAPTER 6

Equipment Usage

Using laboratory equipment correctly reduces the risk of accidents and ensures accurate results.
Using laboratory equipment correctly reduces the risk of accidents and ensures accurate results.

Do :
- Read manuals.
- Receive training on equipment
- Perform regular maintenance checks.

- Read manuals.
- Receive training on equipment.
- Perform regular maintenance checks.

Don't :
- Use equipment without understanding it.
- Skip training.
- Neglect maintenance.

- Use equipment without understanding it.
- Skip training.
- Neglect maintenance.

Image: Proper equipment usage

CHAPTER 5

Chemical Handling and Storage

Proper handling and storage of chemicals prevent spills, reactions, and exposure.

Do : · Use proper containers.
· Label chemicals.
· Store them according to compatibility and safety guidelines.

Don't : · Use inappropriate containers.
· Fail to label chemicals.
· Store incompatible substances together.

Image: Properly labelled and stored chemicals

CHAPTER 7

Hazards and Waste Management

Proper handling and storage of chemicals prevent spills, reactions, and exposure.

Do : · Dispose of waste in colour-coded bins: yellow for biological waste, red for sharps, and specific containers for chemical waste.
· Clearly label general waste bins for non-hazardous materials.

Don't : · Throw all waste into one bin, including mixing biological, chemical, and general waste.

Image: Different bins for biological (yellow), chemical (labelled containers), sharps (red), and general waste.

CHAPTER 8

Control of Substances Hazardous to Health (COSHH)

Definition & Explanation

The Control of Substances Hazardous to Health, or COSHH, regulations are designed to protect people from the dangers of hazardous substances in the workplace. Image: COSHH symbols, laboratory workers handling hazardous substances with proper protection

Key Elements of COSHH

Under COSHH, risk assessments are crucial. Identify hazardous substances, understand the risks, and take appropriate measures to control exposure. This includes using safety data sheets to understand the properties of chemicals, ensuring proper labelling and storage, and using personal protective equipment (PPE) to minimise exposure.

Do : · Conduct risk assessments.
· Use safety data sheets.
· Label and store chemicals properly.
· Use appropriate PPE.

Don't : · Neglect risk assessments.
· Fail to label chemicals.
· Ignore PPE guidelines.

Image: Risk assessments, safety data sheets, PPE usage.

COSHH Compliance

Compliance with COSHH is not just a regulatory requirement but a fundamental aspect of laboratory safety. Regular training and supervision help ensure that all laboratory personnel understand the risks and know how to protect themselves. Proper implementation of COSHH can prevent health issues, ensure a safer working environment, and promote a culture of safety in your laboratory.

Do :
- Ensure regular training and supervision.
- Understand the risks.
- Follow safety protocols.

Don't :
- Ignore training.
- Misunderstand the risks.
- Bypass safety protocols.

Image: Laboratory workers following safety protocols, conducting regular training.

Example: COSHH And Risk Assessment for Handling

Step	Description	Example
1. Identify Hazards.	**Substances:** Sulfuric Acid **Hazard:** Corrosive, can cause severe burn	
2. Assess Risks.	**Likelihood:** Medium (frequent handling in small amounts) **Severity:** High (severe burns). 2.	
3. Implement Controls.	**PPE:** Gloves, lab coat, face shield, **Engineering Controls:** Fume hood **Training:** Proper handling procedures.	
4. Document Findings.	**Documentation:** COSH assessment form for sulfuric and including handling procedures and emergency activities	
5. Review Regularly.	**Review Schedule:** Annually or after any incident (be) **Responsible Person:** lab safety officer	

By following these steps, laboratories can ensure structured and thorough approach to managing health and safety risks, complying with regulations, and maintaining a safe working environment

Steps for COSHH And Risk Assessment

Step	Description	Example
1. Identify Hazards.	Determine what substances procedures or equipment might cause harm	Identifying chemicals like acids, biological agents like bacteria or equipment like autoclaves
2. Assess Risks.	Evaluate the likelihood and severity of harm from identified hazard.	Considering how likely a chemical spill might occur and how severe the injury could be
3. Implement Controls.	Put measures in place to eliminate the risks.	using PPE, installing fume hoods and ensuring proper training for handling hazardous materials

CHAPTER 9

Conducting Risk Assessments

Definition & Explanation

Risk assessments involve identifying hazards, assessing risks, implementing controls, and reviewing the measures regularly

Do : · Follow these steps to ensure a safe working environment. Identifying hazards, assessing risks, and implementing control measures like using PPE, fume hoods, and fire extinguishers are critical steps.

Don't : · Skip this step or neglect safety controls. Skipping risk assessments can result in unanticipated accidents, chemical exposures, or fires.

Image: Step-by-step process: identifying hazards (chemical spills, fire hazards, biological agents), assessing risks (probability and severity), control measures (PPE, fume hoods, fire extinguishers).

CHAPTER 10

Conclusion

By understanding and implementing these guidelines and regulations, including ISO standards, biosafety levels, general safety precautions, and COSHH, we can create a safe and efficient laboratory environment. Let's prioritise safety and make our laboratories a place where scientific discovery can flourish without compromising health and well-being. Thank you for reading, and stay safe in the laboratory!

Reference

Centres for Disease Control and Prevention. (2021). Biosafety

in Microbiological and Biomedical Laboratories (BMBL) 6th

Edition. CDC.

Health and Safety Executive. (2020). Control of Substances

Hazardous to Health (COSHH). HSE.

Health and Safety Executive. (2019). Risk assessment: A brief

guide to controlling risks in the workplace. HSE.

International Organisation for Standardisation. (2021). ISO

Standards. ISO.

About the Author

David Ojowu holds a BSc in Biochemistry and an MSc in Biotechnology from Coventry University, Coventry United Kingdom. With a passion for scientific discovery and a commitment to safety, David Ojowu has spent years both in academia and industry, witnessing firsthand the critical importance of rigorous laboratory health and safety practices.

During his academic journey, David Ojowu identified a significant gap in the availability of comprehensive safety guidelines for laboratory environments. This experience, coupled with his subsequent role as an educator and industry professional, fueled his determination to bridge this gap.

In his professional career, David Ojowu has worked in various research laboratories and industrial settings, where he observed the direct impact of safety protocols on both the efficiency of research and the well-being of laboratory personnel. These experiences reinforced the necessity for a detailed, practical guide on laboratory health and safety.

As a teacher, David Ojowu has been dedicated to equipping students with not only theoretical knowledge but also the practical skills necessary for maintaining safe and productive laboratory environments. He have consistently emphasised the importance of safety in scientific work, inspiring a new generation of scientists to prioritise health and safety in their research endeavours.

"Laboratory Health & Safety: A Comprehensive Guide" is a testament to David Ojowu's commitment to improving safety standards in laboratories worldwide. By providing clear, actionable guidelines and best practices, [he/she/they] hopes to make a lasting impact on the field of laboratory science, ensuring that safety is never compromised in the pursuit of scientific excellence.

In addition to his professional achievements, [Your Name] is also deeply committed to ongoing learning and development within the field, frequently attending and presenting at conferences, and staying current with the latest advancements in laboratory safety protocols.

David Ojowu's dedication to safety and excellence in the laboratory is not just a professional mission but a personal one as well. Through this book, he aims to empower both students and professionals to create safer, more efficient, and more effective laboratory environments.

www.ingramcontent.com/pod-product-compliance
Lightning Source LLC
Chambersburg PA
CBHW051951210526
45474CB00003B/85